# Animal Life Cycles

Written by Susan A. DeStefano

## Table of Contents

# Life is like a circle that goes around and around.

Animals are born. They grow. Then when they are old enough, they have babies, and life begins again. We call this the **life cycle.**

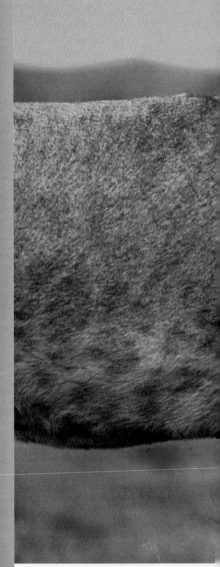

loons

lions

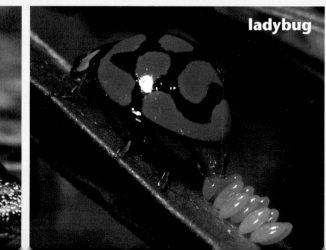

ladybug

polar bears

3

# Life Begins

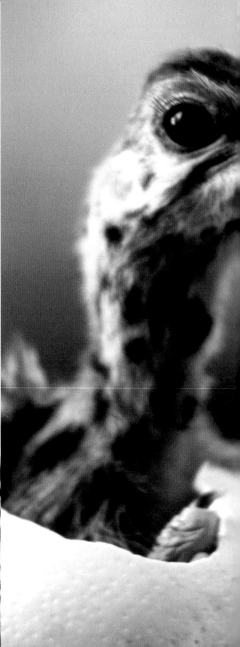

All living things have a life cycle. For most animals, the life cycle begins when the female lays an egg.

When the baby animal is big enough to live in the world, it breaks out of its shell. The egg **hatches**.

tadpoles

4

This mother spider carries her eggs in a silk sac. The eggs will hatch in warm weather.

This caterpillar is breaking out of its egg. In time it will become a butterfly.

Not all animals hatch from eggs. Most **mammal** babies grow inside their mother's body and are born alive. Horses, monkeys, and cats are mammals. Mammal moms take care of their babies after birth. The babies feed on the mother's milk.

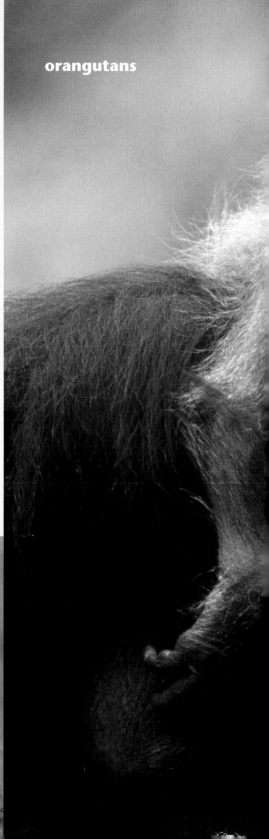
orangutans

kangaroos

Whales are mammals. A baby whale gets milk from its mother.

A bobcat litter is usually two or three kittens.

7

# Growing Up

Most newborn animals look like their parents when they're born. Others grow to look like their parents over time.

Animals take different amounts of time to grow up, or **mature**. Some take only a few weeks, but others take years.

tree frogs

The flamingo chick's feathers turn pink when it's about one year old.

8

A baby giraffe, or calf, is about 6 feet tall at birth.
When it is finished growing, it may be 18 feet tall.

Mammals usually learn to care for themselves from their mothers. They learn to hunt, to keep themselves clean, to find shelter, and to keep safe.

Some animals that hatch from eggs must learn on their own. Their mothers leave just after laying their eggs.

Baby sea turtles hatch on the beach. Then they scurry to find their own way to the water.

ostrich

Mother bears teach their cubs to fish.

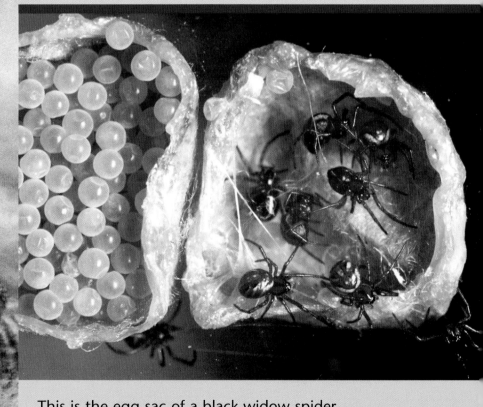

This is the egg sac of a black widow spider.
Babies are hatching inside the sac.

# Moving On

Sooner or later, animals begin to live on their own. Some animals move on just days after birth. Others take months or years. Many mammals stay with their families until they are able to take care of themselves.

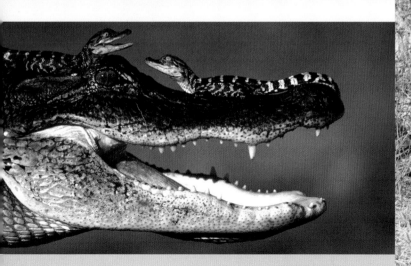

Alligators leave their mother after a year.

**owls**

**elephants**

A baby zebra can't leave its mother until it memorizes the pattern of her stripes.

13

# New Beginnings

Animals continue to grow until they are old enough to **reproduce**—or bring new life into the world. So, although old animals finally die, the birth of each new animal means that the life cycle continues.

| Animal Life Spans | |
| --- | --- |
| house spider | 2 to 4 years |
| fruit fly | 1 month |
| brown bat | 30 years plus |
| bluebird | 5 years |
| cat | 15 years |
| bear | 31 years |
| alligator | 56 years |

penguins

When some penguins are about five years
old, they start looking for a mate.

The more you find out about the life cycle of animals, the more amazed you will be. Pick an animal you really like, and find out more about its life cycle.